All in the family!

Fur, hair and feathers living together

Norm Mort

Hubble & Hattie

Hubble & Hattie

The Hubble & Hattie imprint was launched in 2009, and is named in memory of two very special Westie sisters owned by Veloce's proprietors. Since the first book, many more have been added to the list, all with the same underlying objective: to be of real benefit to the species they cover, at the same time promoting compassion, understanding and respect between all animals (including human ones!) Hubble & Hattie is the home of a range of books that cover all-things animal, produced to the same high quality of content and presentation as our motoring books, and offering the same great value for money.

www.hubbleandhattie.com

First published August 2022 by Veloce Publishing Ltd, Veloce House, Parkway Farm Business Park, Middle Farm Way, Poundbury, Dorchester DT1 3AR, England. Tel 01305 260068/Fax 01305 250479/e-mail info@hubbleandhattie.com/web www.hubbleandhattie.com. ISBN: 978-1-787113-38-1 UPC: 6-36847-01338-7. © Norm Mort & Veloce Publishing Ltd 2022 All rights reserved.

Contents

Preface

Born in Canada, I graduated from the University of Toronto, and soon began writing whilst a teacher-librarian in the public school system. Having grown up with a family dog, I had a dog and two cats shortly after marrying Sandy, a Scottish lass who also loved dogs and cats.

As an automotive writer for magazines and newspapers for over 30 years, I have also written more than a dozen books for Veloce Publishing on historic vehicles, becoming one of Canada's most prolific authors on old cars and trucks.

My first book for Hubble & Hattie was *Dogs on Wheels*, which combined two hobby passions. This, my second animal companion book, focuses on furry, feathered and hairy friends; it certainly will not be my last!

Our son, Andrew, a graduate of the Ontario College of Art and Design (OCAD), has supplied the photographic expertise for all of my books.

This book would not have been possible without the help of dozens of individuals and families,who very kindly shared their stories, experiences and animal companions with us, and so, many, many thanks go to the following wonderful people: Glenda and Wes Myer, Mary Orr, Melanie Parkes, Janice Windsor and Gary Gamble, Cindy and Rick Rogers, Cheryl Douglas, Joan Watson, Shirley and Mike Moumouris and their three children Samantha, Elizabeth and Caroline, Nicole and David West, Zahra, Miguel De Lemos and Ian Nelmes, Glen and Gabby Eustace Donaldson, Veterinarian Doctors Brent and Sue McLaughlin, Kinder Kennels owner Anna Sanford, Richard and Marian Woodlley, Jennifer Cobb, Lee and Lauren Tremblay, Lori Campbell and family, John, Terry Lynn, Jacob and Hannah Phillips, Dee Hazell of Stonefield Eden Farm, Tamara Des Cotes & family, Brittany Boudreau, Enfys Photography, Alice Zhao, Lynn and Jim McPhail, Catherine Mort, and Craig Alexander. I hope I've not missed out anyone!

Special thanks to my loving wife, Sandy, for her nifty ideas, good suggestions, and initial proof-reading. Thanks also to Jude for her ideas, encouragement, and publishing support, along with her dedicated staff.

That's me, Sandy, my wife, and our two Golden Retrievers, Austin I and young Sage, enjoying a day at home in Wellington, Ontario. I am a lifelong auto enthusiast, writer and author, which explains how Austin I got his name; Sage is taken from the rarely heard of French Sage automobile built just after the turn-of-the-century. Ghost (as in Silver Ghost), our cat, has lived up to his name by disappearing!

(Courtesy Michael Lopez)

Introducing our children to pets became a lifelong lesson in love and caring. Here's Sandy with our young son, Andrew, and two of our dogs. Andrew, his wife, Catherine, and their two children are also confirmed animal lovers.

Introduction

Written for all those interested in animal companions, this book will be of specific interest to animal lovers around the world.

According to statistics, the 2022 worldwide dog population was estimated to be 900 million, with 400 million cats.

During Covid-19 more people sought comfort in having a companion animal, with an estimated 23 million US households acquiring one between March 2020 and May 2021 alone. Around 69 million US households have at least one dog, making canines the most popular companion animal there, whilst 45.3 million US households have a cat. As of 2022, America is the world leader in ownership of both dogs and cats.

American statistics in particular do vary from source to source, and from state to state, the reason for which appears to be the fact that the terms 'humane society' and SPCA are generic in the United States. Apparently, shelters labelled under those names are not part of the ASPCA or The Humane Society of the United States. Despite computer technology, no Federal government institution or animal

Cats and dogs in most cases can get along quite well. Some dogs and cats love each other's company; others peaceably co-inhabit, respecting each other's living space, while a few will compete for the position of alpha. (Courtesy Joan Watson)

organization has responsibility for tabulating national statistics for the animal protection movement.

Britain also experienced a rise in companion animal ownership during Covid-19. As of 2022, 12.5 million households in the UK contained a dog (33% of households), with cat ownership coming a close second: 12.2 million households have one (27% of households).

Whether you go for a single animal or, as in this case, a 'double on the rocks,' the love will be intoxicating! (Courtesy Zahra)

This book will also be relevant for and appeal to potential animal owners who aren't sure which animal or animals they would like to give a home to. Via studies and lots of real-life examples, I hope to convince animal lovers that two is company, and three definitely isn't a crowd.

As well, there's a subliminal message that deciding to take on more than one animal will, in so many cases, give another a chance of life – and a longer, happier one, at that. Millions of animals die every day through want of a home: adopting an animal from a humane society or rescue centre will help in this respect.

Heartbreakingly, we can't save them all at the moment, but can do our bit by offering a home and our love. The want-to-be animal companions are eager for a chance; all it takes is us.

To quote my publisher's mission statement: Hubble & Hattie is passionate about animals and their importance and significance in our personal and working lives.

Mixed in with all the facts, findings and opinions in this book are accounts of real people, their animals, and extraordinarily joyful, heart-warming animal family, proving my point that, under most circumstances, sharing our lives with multiple animals will equate to multiple, mutual love, fun and companionship!

It's certainly a case of 'all in the family' in this lovely shot. The Moumouris girls, Samantha, Elizabeth and Caroline, pose for a Yuletide moment with their loving Shih Tzu Poodle Aliki and Collie Sawyer. As mom, Shirley, warmly affirmed, "Our animals are part of the family." (Courtesy the Moumouris family)

It's true: animals are good for you!

Dogs, cats, horses, birds: all become animal companions because of us, and, in many, many ways they make us much better people. Comic superheroes have great strength and fantastic abilities, but they are not really super human. Animals can make us super humans by bringing out the love, kindness and caring we all intrinsically have within us, but all too often don't show enough.

Animal companions give us so much: unstinting companionship; unconditional love; absolute loyalty. Who else greets us as if we've been away forever when, in fact, we've only stepped out for an hour; who else is always, *always* pleased to see us, no matter how we are feeling, cheering us up when we feel down, and reminding us that we are truly loved?

Being greeted with ecstatic enthusiasm by one loving little beastie is wonderful; receiving the same rapturous welcome from two or more must surely be fabulous! Heads cocking side-to-side, our animals patiently listen to our tales of woe, the day's little upsets, and our miseries about aches and pains. They even appear to understand our frustrations

Friendship between two different species is more common than most might think, and sometimes the term 'friendship' is an understatement. Strange though it may seem, real affection between two species is possibie. (Courtesy Rick & Cindy Rogers)

about the traffic, grocery shopping, or the weather (the latter topic being particularly important as it affects walks, being able to sit by their favourite opened window, cuddling in the sunshine on the balcony, or exploring the garden with you).

Yes, two cats, dogs, birds or a combination of species will help keep each other company, but your animals also require human interaction to flourish.

(Courtesy Janice Windsor & Gary Gamble)

Let's talk

Animals thrive on companionship and, although we don't see or hear them chatting away, they do communicate with each other in their own way. And they communicate with us, also, of course: a dog's lick, wagging tail, or bark; a cat's meow or purr, or a bird's nuzzle or chirp, are all clearly responses to being spoken to by their carers, and very often are conveying something specific: a need for food or water; wanting to go for a walk, or to play, or as an alarm about something unsettling.

Some may think that cats and birds are less obviously demonstrative, but carers know (or believe we do), that our feline or avian friend can understand what we say and know what we mean: if not by actually recognising the words, then by our tone of voice. Animals exhibit this understanding and respond in a variety of ways and nuances. As a result of this, we are very often accused of anthropomorphism – attributing human qualities, feelings and characteristics to our animal companions – but sound scientific proof

exists that sentient beings *do* experience very many of the same emotions as us.

Family matters

Sharing our life and home with animals provides us with endless love and companionship, and if you are fortunate enough to have a partner and/or family, all members of your clan will benefit. And, according to current scientific research, when animal and human look into each other's eyes, a hormonal change is triggered that makes us both feel good!

Young children have a natural affiliation with animals, and love to talk and play with their dogs and cats. In return, dogs

Animals and children are natural friends if you have the right combinations. Cats or dogs are the most popular choices, but many other species are worth considering, too.

in particular, tend to instinctively protect 'their' young humans. The strong bond that exists between a child and his or her animal companion is often second only to that they have with their parents.

Studies have discovered that, in many cases, those children who have been raised in a home with animals develop better social skills through increased interaction with the animals. Assuming responsibilities for their animal companions is also an important factor in their social growth and maturity. According to many psychological studies, children with animals are often recognized as having greater empathy, improved self-esteem, and better overall emotional health.

As well, there is medical evidence that spending more time outside with the family dog means that children suffer less from allergies as a result of early exposure to allergens and micro-organisms brought into the house by animals that possibly strengthen the immune system.

As far as historical evidence of living with more than one animal goes, in Germany in 1914 a household consisting of a man, a woman, and two dogs was discovered, dating from around 14,000 years ago in the Palaeolithic period.

Personal footnote (or should that be paw-note?)

Since we married, Sandy and I have always shared our home with three companion animals: sometimes it was two dogs and a cat; other times two cats and a dog. When we lost

What could be nicer than a peaceful, walk on the beach with just two happy dogs playing in the morning sunshine? (Courtesy Adelaide Utman)

Sage as a puppy in 2012 with Austin I. She idolised him from the very start. Sage copied all of Austin's friendly manner and his great habits ... and some of his idiosyncrasies, too!

our Golden Retriever, Austin I, in 2015, we were down to just one animal. Ours was now a far too quiet household ...

All of the animals we've had have been loved and loving companions, but Austin's arrival was special. Our son, Andrew, was off to college, and I had left a profession that had demanded I be away most days each week.

Sage quickly accepted Austin II, and he hero-worshipped her from the start. Out to play together, eating together and cuddling together: Austin II is now learning from Sage.

All in the family

The three of us together – Sandy, Austin and me – generated great joy and love in our home.

With the loss of Austin, our remaining dog, Sage – immediately, and willingly – took on the responsibility of being our centre of affection. As time passed, her dependency on us and unconditional affection made us think twice about adding another dog to the family, or any other kind of animal, come to that.

Always firm believers in the love and joy that results from sharing our home with more than one animal, it was a difficult time for Sandy and I, as we were afraid that Sage would not be happy sharing our love and attention with an additional four-legged friend. Writing this book brought me into contact with so many wonderful people who have happily shared their adoration and sincere feelings for their animals amongst them, and, by the time I had finished all my research and interviews, I knew in my heart I needed another dog. Fortunately, hearing all the wonderful stories I had to tell her, Sandy felt the same way.

We considered alternatives to the Golden Retriever breed, but, like so many dog owners, having found a breed that touched heart and soul, we were unable to stray far.

Whilst we seriously considered rehoming a rescue animal, with Sage being an older dog, we felt she would co-habit better with a puppy.

Looking at puppies once more, we found a litter of Golden Doodles from a breeder very close by, and discovered that the second of the eight puppies had already been named Austin: given my love of all-things automotive, this seemed like fate. We had already decided that he was the puppy for us before we ever laid eyes on him, and so he became Austin II.

Thorough research and practical logic is the intelligent way to decide on your new addition to the family, yet an out-of-the-blue set of circumstances can sometimes intervene. And, as everyone knows, it's never wise to fly in the face of fate. Although definitely not the smartest way to choose a puppy, all has worked out very well for us and Sage.

Choose with the head as well as the heart

Whilst keeping each other company when there's no one else around is a reason for having two animals in the house, some species – such as birds – require company for their health and social welfare.

Bird brains

Canadian bird owners Janice and Gary noted: "Birds, in particular, are very social and flock-orientated creatures who need mates or buddies to keep them company. Birds require a lot of mental and social interaction."

Thus, homing a single cockatiel or parrot might be a mistake that could require more human attention than you are willing or able to provide. Even the presence of another species would help, as birds can suffer from loneliness and depression. Janice went on to explain: "When one bird dies, you are pretty well compelled to get another, or risk the survivor wasting away from a broken heart."

Choosing the right bird or another species should be carefully done. Mary Orr, a keen bird owner, advocate and ornithologist in California who is featured in the next chapter,

While birds love to sing or talk, they also squawk, just as dogs bark and cats yowl. Birds squawk for a variety of reasons, ranging from warning calls to just wanting attention ... and everything in-between!
(Courtesy Mary Orr)

has kept birds all her life, and is an officer in her local exotic bird club, as well as its newsletter editor. She explained why many people are vehemently opposed to a predator/prey mix in a confined setting.

"While the theme of birds and other species together sounds homely and comforting, there is a real dichotomy in the world of bird people. Some believe that 'tame' and 'gentle' animals, such as cats and dogs, can, and will, get along with birds, even though one may be a natural predator, and the other natural prey. Then again, others, like myself, believe that even though humans have domesticated the dog – and the cat, somewhat – their natural instincts can, and have, overtaken them in a heartbeat, depending on the situation. Those ingrained natural instincts far outweigh man's relatively recent attempts at domestication. When this happens, it can result in the injury, maiming, or even death of your beloved bird.

"Yet, make no mistake: a large parrot is a formidable adversary with a bite power of 300lb/psi. A parrot/cockatoo can deliver a grievous bite, and will defend himself to the death. So, it is possible that the other animal could also suffer injury, such as a punctured eyeball or lacerated flesh, etc, despite the predator's size being an apparent advantage."

As for similar avian species co-inhabiting, even cockatiels don't always get along with their own species. Mary said that her birds, Mick, Niles and Coconut, show jealousy towards each other.

When Niles is out of the cage, Mick bounces up and down in an agitated manner; when Mick is out of the cage playing, or Mary's holding him, Coconut throws garbage from the bottom of his cage out onto the carpet.

Niles, the youngest, watches the antics of the two oldest cockatoos, Coconut and Mick, whose posturing consists of stretching out and lifting their crests to full height, hissing, and quick, staccato gestures. Niles finds this psychotic and stressful , and has pulled all of the feathers from his chest, leaving it bare. He has been checked by three avian specialists, who confirmed that the problem was not medical. Many experienced bird behaviourists feel the cause could very well be emotional or psychological. Physically preventing Niles from feather-pulling was facilitated by fitting him with a little sweater.

Said Mary: "Niles accepted wearing the sweater after a few days, although I've seen other parrots who have reduced their protective sweater to unrecognisable, tangled, stringy fuzz."

As for the emotional/psychological problems, these require time, patience and a caring and loving owner to literally smooth ruffled feathers, though often the situation is never completely resolved.

Regarding animals with fur, most cats are quite happy to cuddle up with other felines, or at least co-habit tolerably well. Yet, while they will accept each other, a cat's relationship with her human carer sometimes requires time and often compromise; not to mention a touch of psychology, perhaps.

Pleased to meet you!

Although in most instances good-natured dogs will accept another easygoing canine, this is not always the case. It's possible, over time, to get a dog to adapt (to a point), but it can require a lot of effort – not to mention patience.

Dogs of similar breeds normally get along almost immediately, yet some breeds

Anna Sanford plays with her three Golden Retrievers. Orlagh, from the Netherlands, is 8 years old and still competing in obedience. Eilish is 6 years of age and Aoife-Rose is 3: both are Canadian Grand Champions and hold obedience titles. Orlagh and Eilish are active members of the St John's Ambulance Therapy Dog team in Quinte, Ontario; Aoife-Rose is looking forward to joining them when she is old enough.

are not as willing to tolerate other dogs, particularly if there is a difference in size.

Anna Sanford of Kinder Kennels in eastern Ontario, has been training dogs and teaching dog psychology for over 16 years in Germany, the UK, and Canada, and is a qualified dog trainer with the Animal Behaviour College based in the US, recognized there and in Canada. Anna has years of experience, and has studied and worked in Germany under Dave Mair MBE, as well as being a member of the Canadian Kennel Club, the Golden Retriever Club of Canada, the Golden Retriever Club of Greater Toronto, and the Canadian Association of Professional Pet Dog Trainers.

Anna believes that how a newcomer is introduced to another dog or a pack is of paramount importance, and strongly suggests carrying out the introduction in a neutral environment rather than in the home. And, whilst every dog is different, certain reactions are predictable.

Some breeds are natural pack animals, happy to mix, but others, such as the terrier and guarding breeds, are often liable to be difficult in socialising situations.

Age can also be a factor. Puppies and young dogs are naturally curious, whereas a six- or 7-year-old animal often exhibits little interest. Well balanced dogs are more easily integrated, whereas highly strung canines may exhibit stress or anxiety.

When Anna introduces a new dog into her continually changing kennel pack, she bases it on the dog's usual pace. Slow and steady, she introduces one carefully chosen dog at a time in a separate, large, indoor area. Most often the owner is present to comfort and be a familiar presence for their dog in the new and potential scary environment.

With canines ranging in size from small-medium to a Leonburger and Irish Wolfhound, various methods of introduction are required. Mixed breeds are normally no problem, but purposely-bred 'Designer Dogs' can exhibit uncharacteristic traits. Over-breeding has also created problems, and altered the traditional nature of breeds in some cases.

Anna advises that if a potentially problematic situation arises when dogs are mixing, to step in immediately, and if a loud 'HEY!' doesn't work, use a water spray or even a hose to defuse the situation (although never the latter method when introducing a new dog with a newly-rescued dog, or one who is already nervous or anxious).

Very different breeds can get along famously, depending on characteristics and temperament; equally important is disposition. Through years of experience teaching canine behaviour and psychology, Anna Sanford of Kinder Kennels knows her breeds, and can spot the signs that help discern an individual dog's possible reactions to other dogs. Some of her dog training experience was acquired during the time she was an officer in the British military.

It's important to understand why conflict occurs. Although jealousy can be a factor, it's also the case that many dogs are protective of others, and will come to their defence. Interestingly, often a younger dog will try to protect an older one.

Each breed tends to have different issues, Anna says. For example, Great Danes can feel stressed when separated from their family, refuse to eat, and develop subsequent problems. A Mastiff bred to guard is instinctively protective and likes his own space. Herding dogs are bred to work and not socialise with other dogs, which can cause problems. Poodles, when over-bred, can be grumpy, and Poodle crosses can exhibit unstable behavioural traits in addition to suffering numerous health issues.

Rescue me

Introducing a new dog into a home with already-established animals can be an easy transition or fraught with problems; maybe due to breed or, in some cases, the newcomer's past experiences. Taking on a dog from a rescue centre can be a very rewarding and worthwhile experience, although not always.

One large dog rescue turned out to be a terrific family animal and tolerated a cat, but

A Yorkie, Crested Mix, a Border Collie and an Aussie Shepherd with a Double Merle condition. As Zahra explained: "A Double Merle is a genetic condition/colouring name. The Merle gives this patterned colour that's very popular, but, sadly, when two Merles are mated, up to half of the pups can be deaf and blind. Backyard breeders and puppy mills with Double Merles sometimes drown the deaf/blind puppies. My sister's pup was lucky to end up in rescue." (Courtesy Zahra)

absolutely hated small dogs. After being forced to survive in the wild for months on very little food, another dog – though gentle as a lamb with people – had no problem taking on dogs large and small, or wildlife, because of his earlier experience of foraging to survive.

Sad beginnings in life due to poor treatment by people psychologically scars some animals, requiring huge amounts of love and patience to overcome. It is essential to know as much as possible about your rescue animal's past life and experiences before you commit.

Re-homes and rescues are fine alternatives to the traditional breeder that all dog lovers should encourage. It is also an appropriate and responsible substitute for many who find it difficult to get through the puppy stage with a dog, or for anyone who loves dogs and would, for a variety of reasons, be better off with a mature dog or dogs.

Zahra is a firm believer in re-homing rescues, and has had years of experience finding appropriate homes for them through volunteering her services at local shelters. Her own two dogs are from shelters, and she couldn't be happier with her dynamic duo. "Technically, I rescued them but, trust me, they

rescue me every day, and bring such joy and love to my life. I don't know what I would do without them."

Zahra tries to help potential re-homers of rescue dogs think of what life will be like now, *and* ten years down the line, should they take on a dog. She asks what they will do with their dog; how often and for how long will their dog be alone; how do they plan on training their dog; what doggie traits are deal-breakers for them? She suggests to many would-be adopters that they should consider dog-minding rather than taking on a dog of their own, as their lifestyles aren't conducive to a new addition becoming a well-balanced and happy companion.

Smaller dogs require everything that larger dogs do, of course, though size makes a difference when it comes to the amount of food they eat, exercise they require, and ease of tackling certain tasks, such as grooming and bathing. Owner Zahra noted, "The first time I saw the dual kitchen sink in our county home I thought wow, it's a dual doggy bath! My real estate agent wasn't sure if I was kidding. Naturally, the dogs don't always seem as delighted as I am with the baths. Go figure." (Courtesy Zahra)

As Zahra further explains: "An unhappy dog makes for an unhappy human companion. For them, I offer play-dates."

Double-trouble?

Whilst there are many benefits to having more than one dog there are also drawbacks, and expense is just one. Then there's training.

Zahra says:, "It's harder to train two dogs simultaneously, so you need the time, patience and money to be able to provide individual training. It can also be a challenge if one dog has behavioural issues, as there's a chance the other dog may learn these, too. Two dogs can also be louder than one, especially when they play or are naturally guarding their home. They can become jealous of one another which, if not nipped in the bud right away, can escalate into aggression towards each other.

"Having two dogs is no reason for not properly exercising and socialising them. When out with two dogs, it can be hard to monitor both and keep them safe and interacting well with others. Remember that if you don't socialise your dogs with other dogs, they won't know what to do when they meet other canines, as they are bound to do.

"And travelling with two dogs can be more difficult, dependent on type of transportation, size of the dogs, and whether you are alone. If a commercial carrier is required then it's their rules."

Although she calls her home Stonefield Eden Farm, Dee Hazell's philosophy is quite different to that of a traditional farmer, as its aim is to combine her love of animals and her passion for wool and knitting.

All of her alpacas and the two llamas have names, and are regarded as companion animals. Although her alpacas provide the wool necessary for her knitting business, they are part of the family. Dee houses her young alpacas and their mothers and aunties in the century-old bank barn. To keep everyone safe, two livestock guardian dogs (LGD) live full time with the alpacas and a variety of poultry, including laying hens, a few roosters, two ducks and a goose. The LGDs protect the alpacas and poultry from coyotes and wandering dogs.

Dee also has two horses, one of whom is helping her learn to ride, whilst the other helps her mount, and also an inside dog and numerous cats who have few responsibilities other than providing love and companionship. Chance, the house dog, is both a doorbell and an alarm system. Dee recently acquired a few guinea fowl who alert everyone when anything or anyone comes onto the property.

Chickens and a rescued goose have happy lives with Dee, completing this picture of bucolic bliss.

All in the family

Farm life?

If your heart's desire is a barnyard full of a wide array of livestock and fowl, don't forget to take into account the many serious financial considerations that this will entail, ranging from specialised housing to vet bills that go well beyond those of caring for a cat, dog or bird.

Dee Hazell of Stonefield Eden Farm bought her place in the country, and found the determination – despite the lack of a farming background – to live her rural dream. She always felt she should have been raised on a farm, and relished childhood vacations spent at her aunt and uncle's farm in Prince Edward County, and the Glenbrook Vacation Farm in the Eastern Townships in Quebec.

Dee has a profound love of animals, and a true desire to take care of them and enjoy their companionship in a simple and quiet rural setting.

Many others feel the same, and some I have met in my travels have kept chickens to enjoy fresh eggs, but became so attached to their birds that they have given this up and become vegan. Traditional country and farming folk may well consider this stance foolhardy: Sandy and I prefer to view it as a love for all creatures – great and small.

Visit Hubble and Hattie on the web:
www.hubbleandhattie.com • www.hubbleandhattie.blogspot.co.uk • Details of all books
• Special offers • Newsletter • New book news

20

Birds of a feather

Mary Orr with Niles, Coconut and Mick

Mary Orr of California is a bird-lover, and has three Umbrella Cockatoos who are her companions. Said Mary: "I had dreamed of having a Cockatoo for a long time, having had small parrots previously. One day I decided it was time. I found little 5-month-old Niles at a respected bird shop in town and brought him home. I intended for him to be an 'only child.'

As (fortunately) so often happens with bird and other animal owners, however, one often leads to two, and a second Umbrella Cockatoo – Coconut – was adopted a year later to save him from a pet shop that had a dubious reputation.

Recalled Mary: "I saw a cockatoo in the shop and was drawn to him immediately. I asked if I could pet him and was warned, 'Okay, but watch out, he's pretty mean.' I could see that he had absolutely no toys or anything to occupy his time and mind, which is wrong for an intelligent animal."

What happened next says as much about animals and birds as it does about

As bird-lover Mary Orr pointed out,
"Although I realise the dog/cat population is
overwhelming, there are many bird owners
out there, too. But, in the pet section of stores,
nearly every product is for a dog or a cat.
Extreme discrimination!" (Courtesy Mary Orr)

people. "I approached the cage and slowly put my hand in, speaking softly to him, and very carefully watching his reaction and body language. He moved slowly as well, and proceeded to lay his little head in the palm of my hand. I was a goner!"

The next day when she collected Coconut, Mary says he bounced up and down with happiness on the way out of the shop!

Mick was a friend's Cockatoo who visited occasionally, but when his owner moved, Mary ultimately ended up with Mick staying permanently.

Mary recalled: "When Coconut and Mick first met, they began to go claw-to-claw, and had to be separated, as each was trying to assert his position as alpha Cockatoo."

At first, young Niles was very friendly with both other birds, but, as time went by, he eventually aligned with Coconut, and now eschews Mick, too. Sadly, to this day, Coconut

and Mick are not friendly, and very jealous of each other. As a result, Niles and Coconut will play together, but neither of them will play with Mick.

Mary Orr's bond with her three male Umbrella Cockatoos has continued for almost a quarter-century, with Niles now 24 years old, Coconut 32 years and Mick 30. Yet, while the three enrich Mary's life, and are endlessly entertaining and amusing, as with all animals, there are downsides.

"They can be very loud" Mary said "and, because of their natural instinct to chew things, destructive, too. Therefore, they have to be supervised when out of their cages."

And although it is a very rare occurrence, a Cockatoo's bite can be very painful, as Mary has reason to know. "Misting the birds with water conditions their skin and feathers, but it doesn't mean they like having it done. Once, I was misting Coconut, and he became irritated by my persistence, and gently took my hand in his beak and moved it away. I didn't take the hint, however, and continued misting him. Twice more he did this but still I continued, so then he took my hand hard in his beak, applying some considerable pressure, and piercing the skin. After a loud 'Owww!' and some profanity from me, I called Coconut

Well, sometimes that old adage two's company, three's a crowd can hold true. While cockatoos Coconut and Niles are affectionate with one another, Mick is the odd man out. Just like some siblings are jealous of each other, or feel they should be dominant, animals – and, in this case, birds, are sometimes the same.

Like any mom or dad who loves all their kids, you cope and try to keep everyone happy!
(Courtesy Mary Orr)

'bad bird' several times, which wasn't really fair because he had warned me! (I don't think he really knew his own strength, and I certainly don't believe he meant to hurt me: he was just trying to be emphatic.)

"I pointed to my hand and said: 'Bad!' and Coconut turned his head away. He tried to apologise by making soft, cute little noises whilst watching me, but I ignored him for hours, despite him repeating this tactic. I heard banging sounds and discovered he was tearing up his favourite toy and throwing the pieces to the cage floor! Clearly, he was expressing frustration and anger. After that we made up and everything was fine – well, except for my swollen hand!"

Mary concluded by saying: "Parrot-type birds are as intelligent as a 5-year old (per tests/experiments), and many are very affectionate and cuddly, like a dog. Some even mimic human speech. They are social, entertaining, interactive, and have huge personalities. They are some of the world's most beautiful creatures – not that I'm biased, of course!"

Doctors in the house!

Brent and Sue McLaughlin are both veterinarians living in Beaverton, Ontario, and share their home with two domestic longhair (DLH) cats – neutered, of course.

Why two? Well, one for each daughter! Blackie is a 4-year-old male, while Beanie is a 2-year-old female.

Despite Brent and Sue being veterinarians who must deal with the sadness of euthanizing animals on all too regular basis, it never gets any easier, and both felt the one drawback of owning an animal was the grief felt over their inevitable death. Yet, as Brent

The McLaughlin family cats, Blackie and Beanie, get cosy in the clothes basket!
(Courtesy the McLaughlin family)

noted: "Our cats are a continuous source of comfort and humour, and a constant reminder of God's creative beauty."

When it came to choosing their cats, the McLaughlins were truly hands-on.

Sue neutered Blackie when he appeared on the neuter roster while she was working at the Georgian College in the Veterinarian Technician program. Blackie had been found in a garbage dump as a kitten and taken to a shelter. Beanie was a gift; adopted by the McLaughlins shortly after they lost their long-time cat.

Brent said Beanie is known as a 'purr machine.' "She is a wonderful little cat who came to us very thin, but has definitely made up for that with her voracious appetite."

Apparently, when Blackie and Beanie met there was some hissing and distance before they agreed to disagree and maintain respective boundaries ... but that soon changed.

As a DVM (Doctor of Veterinary Medicine) Brent recalled: "When we first got Beanie she was not spayed, and so her first heat became a bonding time for the two cats. But, since Blackie was neutered, it appeared that he found this once 'elusive, nasty cat' quite a nuisance when she would not leave him alone while in heat.

"And, since they are induced ovulators [the act of breeding stimulates the release of eggs from the ovaries] the behaviour went on for days [cats will stay in heat for 7-10 days on average but, if they are bred, will ovulate and come out of heat 1-2 days after ovulation] until the relevant organs were removed by Doctor Sue."

Ever since, to all intents and purposes, Blackie and Beanie have enjoyed each other's company and playing together.

Brent added: "They love to run around the house; thankfully, we have old, well-worn hardwood floors, so when they skid there is virtually no damage."

Country life for Glenda, Wes, Bert and Ernie

Glenda and Wes Meyer live in the country in south eastern Ontario with their two 10-year-old male English Springer Spaniels: brown-and-white Ernie and black-and-white Bert.

No, they're not named after those two well-known Muppets of Sesame Street fame, but rather the equally loved characters from

Playing in the snow is a favourite pastime of Bert and Ernie, but being warm and snuggled together inside looking out is a bit of alright, too!
(Courtesy G & W Meyer)

the classic Christmas film *It's a Wonderful Life*, in which Ernie and Bert are the amiable taxi driver and policeman friends of George Bailey.

Glenda and Wes chose English Springer Spaniels out of tradition. Glenda explained: "We had lost another Springer – Little Moo – and, after a year without a Springer in our lives, an advertisement for puppies caught our eye. We made the drive to Cardinal, Ontario, and met them, along with their ten siblings. It was love at first sight."

Bert and Ernie celebrate their second birthday with a big party and matching neckerchiefs!
(Courtesy G & W Meyer)

All told, the Meyers have previously adopted four Springers and now these two: 'The Boys' as they affectionately call them.

Explained the Meyers: "We had intended on having only one puppy. A picture of Bert had caught our attention, but when we arrived they were all so wonderful, and Ernie just couldn't be left behind. And what a wonderful decision that was. They are great companions: very loving and sensitive. There isn't anything they enjoy more than a snuggle on the couch. They aren't the best guard dogs since they love everyone, but we wouldn't be without them."

The only thing that could be considered remotely negative about these two lads is the unfortunate fact that they are strangely attracted to porcupines ... ouch!

"Especially Ernie," said Wes, "so when darkness falls they can't be left outside alone."

It was also mentioned that Ernie enjoys his food and quickly polishes off his meals, whilst Bert, on the other hand, is a slow eater who savours every bite. Not surprisingly, this can cause some friction when Ernie has finished his meal and tries to sneak up on Bert's bowl. Though maybe he simply intends to hurry Bert along so that they can play.

Glenda and Wes established a couple of play routines right from the start. "Since they were puppies, one carpet in the house has

English Springer Spaniels are quick learners; in no time at all they can be running the place. "Will that be cash or card?" asks Bert; "Will you need a bag?" enquires Ernie. (Courtesy G & W Meyer)

been designated as the 'Playground,' and at ten years of age, they still go there to have a roll together and chew on their toys ... especially the loud squeaky ones!" laughed Wes. "They both love the snow, and winter will find them out playing, finding the highest snow bank to perch on and ultimately arriving at the door, waiting to come in for a biscuit looking like giant snowballs."

What does it mean to Glenda and Wes Meyer to have these two Springers in their lives? "We always have someone [meaning Bert and Ernie] at home who loves us unconditionally."

Glen and Eustace (aka Gabby) Donaldson, plus Westies

Burlington, Ontario residents Glen and Eustace (Gabby) Donaldson have always driven and enjoyed vintage post-war British two-seater

sports cars. For over three decades they've owned a 1978 Triumph Spitfire 1500, and had other classics, too, including a Triumph TR3A, and a TR7 for many years, and have added to their stable two 1985 Morgans: a Plus 8 and a four-seater 4/4, the latter added to the collection as it was a necessity: it had a backseat, unlike the Plus 8.

A highly practical acquisition, the 4/4 could easily accommodate their two beloved West Highland White Terriers on touring events organized by the Morgan Sports Car Club of Canada (I mean, come on, Glen is President of

Dogs who love cars and top down motoring love to go out even in the coldest weather; especially when it's in a vintage Morgan! (Courtesy G & G E Donaldson)

*"Okay, what are we doing, eh, eh?
Going out in the Morgan? Walking? Hangin'
out watching TV?* (Courtesy G & G E Donaldson)

the club and has to set a certain standard!). The Westies have become enthusiastic members of the club, too, and seem to enjoy nothing better than a ride in the Mog.

Dalwhinnie – Whinnie for short – is 7 years old, while her male co-driver, Macallan,

is five. Both are named after favourite Scottish single malt whiskies. (Hmmm ... how about having a cat named McEwan as a beer (or dog) chaser!).

Glen and Gabby adopted both West Highland pups when each was three months old. Winnie was first; Macallan came from the same breeder and is actually Winnie's nephew – that's definitely all in the family!).

Glen remembered: "The first two days Whinnie gave Macallan that smiley, growling face, but, by day three, they had started to play together. Now, five years later, they are inseparable. They keep each other company and love to play."

Most dogs love routine and Whinnie and Macallan are no exception. Said Glen: "Both will start the play, which usually takes place post-morning walk, or sometimes when it is our relax-downstairs-and-watch TV time, when, instead, it's 'Westie Wrestling' time!"

Of the two, Whinnie is more terrier-like, chasing any wildlife foolish enough to trespass into the backyard. Macallan joins in the hunt,

Going out with Gabby for a walk on a nice day also means a spin in the Morgan. West Highland bliss!
(Courtesy G & G E Donaldson)

but being rather laid-back, soon decides: "Oh, wait, I think I need a cuddle with Gabby."

As far as Glen and Gabby are concerned, having two dogs are no more trouble than one, and besides: "Westies are very active terriers, and keep you motivated to get out, walk, and enjoy life."

Miguel De Lemos, Ian Nelmes and their Schapendoes

Miguel De Lemos and Ian Nelmes own Mirazule, a new boutique hotel situated on the shores of Lake Ontario in picturesque Prince Edward County, Ontario. The pair's stunning, ultra-modern living accommodation is shared with two very cuddly and friendly male Dutch Sheepdogs; more officially known as Schapendoes.

The brown-and-white coloured Drambuie (Buie) is 12 years old, and 7-year-old black-and-grey, or Blue Roan, Razzmatazz (Razz) joined the family after Spats, the first Schapendoe the pair had, died of cancer in 2012.

Schapendoes are not a common breed, despite dating back several centuries. Originally trained as herding dogs, the breed almost became extinct in the late 1930s. Valued for their endurance, intelligence and loving nature, these easy-to-care-for dogs slowly began to flourish again following WWII.

When Miguel and Ian lived in London they both worked all day. They wanted animals in their lives, though, so gave a home to two cats. The thinking behind getting two was that the cats could keep each other company, and could be left to their own devices during the day more easily than could a dog.

Situations change, however, and often bring new possibilities: in the new millennium

Ian and Miguel moved to France. As Miguel explained: "In 2001 the change in our lifestyle permitted us to have a dog. We researched a particular breed, but discovered that its temperament wasn't ideally suited to a business that involved welcoming strangers. We then came across the Schapendoes breed, which is actually known as being very sociable and friendly: this suited our hotel business perfectly!"

Miguel and Ian located a breeder in the South of France and waited for a litter to be born. They eventually got their first Schapendoes (Spats) in the summer of 2002. Delighted with Spats, another Schapendoes (Buie) joined the family in 2008.

The hotel's location near a river was well suited for the two energetic Schapendoes, as

Razz was introduced to Buie in the garden of Miguel and Ian's boutique hotel in France. It was friendship at first sight, and the two were soon having fun playing and running together.

(Courtesy Miguel De Lemos & Ian Nelmes)

Schapendoes seem to like everybody; Buie and Razz even like cats! Here they are, calmly greeting their 'cousin,' Charlie, in Barrie, Ontario. Charlie seems somewhat less impressed ...

(Courtesy Miguel De Lemos & IanNelmes)

Miguel noted, "In spite of Razz wanting to be top dog, he is very clingy, and will always be the one to curl up as close to Buie as he can. He often uses Buie's rear quarters as a pillow!"

(Courtesy Miguel De Lemos & Ian Nelmes)

Miguel and Ian could walk them for miles along the tow-path. The two dogs could also run off-lead without danger.

"With the loss of Spats and both the cats, who had died of old age, poor Buie was on his own," Miguel explained. "That didn't seem fair, so, even though we were still getting over the loss of one beloved animal, we decided to get another dog: not only to keep Buie company, but also to bring him out of himself." (This was an extremely insightful and caring call by Ian and Miguel, as people often tend to focus on *their own* loss, rather than how other animals in the family are coping with the sudden loss of a lifetime companion.)

A further observation by these caring owners was the fact that Spats had always been the leader of the two, always running ahead whilst Buie was happy just to run with him.

Miguel reminisced: "After Spats died, Buie would stay beside us on walks, never running unless we ran with him. We realised that we needed the new dog to have a more dominant personality; a leader who Buie would follow. We explained all of this to the breeder and she chose Razz for us."

Upon meeting, the two Schapendoes immediately enjoyed being together.

As Miguel explained: "Buie is a very

Zen, happy dog and nothing fazes him. Being so easygoing, he accepted Razz, and they got on from the first moment. Razz has a more forthright personality, and Buie has always been happy to let him be top dog."

Having two dogs in a household means there's a possibility that jealousy can develop if one feels the other is getting more attention or affection, and Miguel and Ian are well aware of this. "We've always made sure that we give equal time to all of our animals, and let each win games in turn; that is when we're not winning, of course! I'm not sure if this is the reason, but whilst Razz is pushy and plays to win, there comes a point where he in turn takes a step back and lets Buie be the victor. He still gets carried away with an over-abundance of youthful exuberance, but Buie holds his own ...!"

In conclusion Miguel added: "I hate to draw parallels between companion animals and children, but they are undoubtedly a repository for all the love we have inside us. I cannot imagine loving anyone more, or feeling more fiercely protective of them."

It was obvious that both Miguel and Ian felt the same way about their beloved Schapendoes. "We are besotted with them. After Spats, we wouldn't consider another breed, and Buie and Razz have reaffirmed that."

David and Nicole West and their Wheatons

Now retired, David and Nicole West, along with their two Soft-Coated Wheaton Terriers, divide

Buie and Razz are eager to be on the go, be it in the car, for a run or a stroll in the garden, as long as they're with Ian and Miguel.
(Courtesy Miguel De Lemos & Ian Nelmes)

"It's beginning to look a lot like Christmas ..." And, what's Christmas without favourite puppies to help you celebrate? "I wonder what Nikki, our godmother, got us?" ponders Buie. "It'll be good, whatever it is!" barks Razz!
(Courtesy Miguel De Lemos & Ian Nelmes)

their time between Toronto and Florida. Kelsey, the West's first Soft-Coated Wheaton, was joined by a second, Kylie, in 2013.

At that time both of the West's sons, Gareth and Graham, were home, and thus the dogs were rarely, if ever, left alone. As children have the habit of doing, the boys grew up and moved out, leaving David and Nicole and the two Wheatons, but when Kelsey died in 2017, as Nicole put it: "One dog and two humans were not enough to be considered a pack." She adds: "David considered how many years he may have left and thought that, if a Wheaten Terrier's lifespan is 12 to 15 years, we should get another one that year. He also considers that the dogs are better company than many human beings."

Kira joined the pack soon after. Some might have reservations about how a puppy

and a mature 6-year-old dog might get along, but, from my own experiences and that of others, the age difference does not appear to be an issue.

Nicole and David left the breeder with Kira, and arriving at the house put a leash on Kira to introduce her to the backyard. Nicole explained: "Kylie was with our son, and he soon brought her home and around back to meet Kira. Their first greeting was on our patio. Kira immediately decided that Kylie was there for her to play with, and proceeded to jump on her new playmate. Kylie was polite and very tolerant, though tried to keep her distance for the first few days."

Being a pup, Kira was very persistent. Said Nicole: "Kylie would look at us askance, as though we should give her instructions on how to behave around this new, frisky pup. As

After the two Wheatons became used to each other, the Wests found that, "Initially, most of playtime involved Kira jumping onto Kylie or grabbing her by the hair." (Courtesy David & Nicole West)

An older dog can teach a young one new tricks. "Come along, quickly now! I'm going to show you the best place to hide a slipper. It's great fun!" (Courtesy David & Nicole West)

well, Kylie would sometimes give us a look as though she was pleading for us to intervene when Kira was being annoying."

Yet, within a week of Kira's arrival, Kylie had realised that it was okay to play with the puppy.

"It took several weeks before she would discipline Kira if she got too rough or annoying," explained Nicole. "Still, it wasn't long before Kylie accepted that Kira was now part of the pack."

The West's feel Kira will eventually be the more dominant of the pair; Nicole pointed out: "I always feed Kylie first, and she is always the first to be given treats. Kylie is still stronger than 6-month-old Kira, but she puts up with a lot from her younger pack-mate."

And, as pack-mates, the Wests found it was usually Kira who encouraged Kylie to play, although sometimes Kylie will take the initiative. That's certainly a sign that they enjoy

While sleeping in a chair or lyng on a bed is nice, sometimes a different perch is more of an attraction. Said Nicole, "I walked into the living room to find Kylie lying in her usual daytime place on the sofa, but with Kira perched on her back!"
(Courtesy David & Nicole West)

playing together.

Owners are aware that two dogs will keep each other company, but there is often far more to their relationship. Said Nicole,

"Kira had been playing non-stop with another puppy when she noticed that Kylie was just sitting by herself. In the middle of her

Nicole noted, "Kira was like a Tasmanian devil in the garden. At 3 months old she would tear around the yard, flattening Hostas and Irises, and taking flying leaps into all the other flowers, whilst Kylie looked on, bemused. Kylie wasn't destructive in the garden, but soon both of them were wrestling amidst the flattened plants, proving that you can teach an old dog new tricks! (Courtesy David & Nicole West)

Escaping to higher ground ensures napping won't be interrupted by a little sister.
(Courtesy David & Nicole West)

wild playing, Kira picked up a toy, went over to Kylie and offered it to her. She repeated it with a second toy. Seems like she wanted Kylie to have a good time, too. Either that, or she just wanted Kylie to know that she hadn't forgotten about her."

While Nicole and David admit there are drawbacks to having dogs, "... arrangements when travelling; interrupted sleep from sharing our bed; a dog needing to go out in the middle of the night; dirty paw-prints on floors, carpets and sofas, and the Wheaton Velcro-like hair to which bits of stuff cling, and are brought from the garden into the house," David and Nicole wholeheartedly agree it's worth it: "Having Kylie and Kira means that every day we have two precious girls to love us and give added purpose to our lives. With Kylie and Kira around, I feel safe and secure when David is away, and it's great having them hang out with us when we are working in the garden. Having dogs brings warmth and constant companionship, while their antics are funny and entertaining. We love when they cuddle with us. They give us a lot of love and affection, and we have them to love in return."

Rubbing along together

Shirley and Mike, Samantha, Elizabeth and Caroline Moumouris, with Sawyer, Aliki, Blue, Amber, and Petey

Toronto, Ontario residents Shirley and Mike Moumouris and their three children, Samantha, Elizabeth and Caroline, have always had dogs and a Conure Parrot, too, named Amber. As Shirley wisely points out: "Having three children meant having different animals, according to their likes, such as dogs, birds, fish, a gerbil and a mouse."

Originally, the Moumouris had one dog at a time. As Shirley explained it: "We tried introducing other dogs, but they never seemed to match in temperament, and became too much work as the kids were little and we found we could not handle more than one dog." (Hence, the birds, gerbil, fish and mouse.)

As the children got older, the family decided they would like an additional dog, and thus Aliki joined them. Very rarely called by her given name as it sounded too formal, their four-year-old female Shih Tzu Poodle answers to Scooby, Minou, Liki, Boogie and Woogie, and

When little Aliki met Sawyer, recalled Shirley, "Aliki took comfort in Sawyer, and always rested on his tail, or tucked in very close to his body."

(Courtesy the Moumouris family)

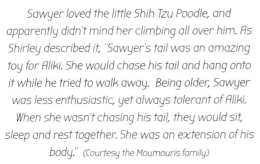

Sawyer loved the little Shih Tzu Poodle, and apparently didn't mind her climbing all over him. As Shirley described it, "Sawyer's tail was an amazing toy for Aliki. She would chase his tail and hang onto it while he tried to walk away. Being older, Sawyer was less enthusiastic, yet always tolerant of Aliki. When she wasn't chasing his tail, they would sit, sleep and rest together. She was an extension of his body." (Courtesy the Moumouris family)

Shirley spoke about an 'all in the family' moment. "Often, when we are all sitting on the floor playing with the dogs, Amber will climb out of her cage and try to join in. The dogs find her very entertaining as she will chase them around. There have been times when Amber has climbed onto their fur, and the dog takes off with her attached. It's funny afterwards, but not at that moment as things can get very loud, very quickly." (Courtesy the Moumouris family)

came on-board as a canine companion for the Moumouris' existing dog, Sawyer the Collie.

Previous to these two pups, the family had had Pooper, followed by Chelsea, both of whom were Collies. Shirley recalls: "When each of our beloved dogs died, the void had been so great that when 12-year-old Sawyer was on his last days, my daughter, Caroline, made a point of having another dog to not only assist us with his passing, but also Aliki who was so attached

to him." [That is very wise and perceptive advice, applicable to all animal carers in a similar position.]

Introduction of the new addition took place in the yard at the Moumouris home. Aliki had been adopted when a year old, and immediately took to Sawyer who was seven. Young Collie pup Blue was just eight weeks old when he was introduced to Sawyer and Aliki.

The meeting between the three went

Thinking back, Shirley remarked, "Aliki would run circles around Sawyer, and he would just watch her. Yet, Aliki and Sawyer were never far apart in all they did: she seemed to be an extension of him. When Sawyer was ill, Alilki would not move away from him." (Courtesy the Moumouris family)

"On meeting the new kid, other than curiosity there weren't any issues in the beginning," recollected Shirley. "However, as Blue has gotten bigger, Aliki does not like him near her in the house, though doesn't seem to mind him outside if they get to chase each other. Indoors, she growls at him, but he thinks it's a game and continues to taunt her." (Courtesy the Moumouris family)

well, without any issues and, as Shirley noted: "Sawyer was quite sick when Blue came to us. He sniffed him quite a bit and seemed to approve of him. All three played outside, and the next day our beloved Sawyer passed away.

"My daughter, Elizabeth, posted that Sawyer had realised it was okay for him to go because he saw that Aliki would not be alone."

At that point Blue and Aliki were the same height. Over the ensuing months Aliki remained small, of course, whilst Blue grew quite a bit bigger.

Aliki is apparently not fond of his agility in the house, whereas outside she will run with him.

In fact, the family soon discovered that Aliki and Blue love to tease and chase each other, though Aliki likes this to happen in the backyard, not in the house, and certainly not near her bed.

When Sawyer was alive he and Aliki would happily eat together, but Blue and Aliki

Shirley described a typical family work day. "Our dogs know ahead of time when one of us is coming home. At around 5:30pm they line up on the porch, even though our car is not in sight. Within a few minutes, however, one of us arrives. It's quite curious, but it happens without fail."

Despite the fact that Aliki isn't as fond of Blue as she was of Sawyer, these two know they have a common bond in their love of the Moumouris family. Here they sit apart, yet together, awaiting the arrival of one of their pack. Shirley Moumouris described it best: "Coming home to these two, waiting on the porch, tails wagging, whatever has happened that day is not important any more: all that matters is the love pouring from them as if you are the most important thing in the world."

(Courtesy the Moumouris family)

have to be served in separate rooms as, with him, Aliki is very territorial about her space. And dog walks with Sawyer and Aliki could be a problem at times, as neither dog would walk alone: they were the best of friends, even on their outings. As Shirley explained: "As soon as one realised they were alone, they would stop and refuse to continue. Now, Blue will walk alone, but still not Aliki."

When I asked Shirley what she thought the benefits of multiple animals in the household are, she replied: "More than one dog has meant companionship for each of them, whether in the house when we're not there or outside playing in the yard. Also companionship for each child according to their temperament."

Of course, there are the usual drawbacks, such as travel limitations, and often the expensive, but always necessary, vet bills.

A further drawback the Moumouris encountered is dog-sitting. It is always usually easy to leave a single dog with a neighbour or family member, but not two or more. And, as Shirley correctly pointed out: "A little dog doesn't take up much room in the car, but a big guy can overcrowd the space."

Certainly, as the Moumouris family found, going on holiday and hiring a house/animal-sitter so that the dogs can stay comfortably in their own home, have the same routine, and not be subject to increased loneliness and confusion, can be an expensive proposition. However, as much as they sometimes flinch when paying the bills, they gladly do this over and over for their loving dogs.

Shirley concluded with: "The animals, especially the dogs, are the first ones to greet

Having more than one dog can be taxing at times. For example, if you pet one, you must pet the other, too. "Dogs can be jealous of each other, when one gets a treat or is petted, say. You have to ensure they each receive the same treatment." said Shirley.
(Courtesy the Moumouris family)

us and the first ones we greet. I notice when one of us is upset, we usually find ourselves sitting with the dogs. Recently, I overheard my daughter, Samantha, say, 'Even when I am home alone, I'm not lonely. The dogs are with me.' The animals add laughter, love, companionship, conversation and therapy. Each brings their own sound, cuteness and mood."

She added: "My daughter, Caroline, has now moved out with Blue, and she finds herself singing to him in the same way that Mike would sing to the dogs. She told me: 'Now I realise why dad would sing to the dogs: the love in their eyes when you do is priceless.'"

As the only dog Aliki seemed lonely,

and Mike was even lonelier. Said Shirley, 'About five months after Sawyer had died we brought home a four month old English Springer, who we named Petey. His abundant energy has kept us fit with walks, runs and dog park visits each day. Blue visits weekly, and, with three dogs, the house feels like a zoo sometimes. We love it!

Richard and Marian Woodley with Nolly and Riley

The Woodley family live in a small village in eastern Ontario. Richard Woodley was a township clerk, whilst Marian is an accountant.

Long-time cat owners, the pair decided to have a dog, and a Wheaton Terrier – being smaller and hypo-allergenic – was the best choice for them, they felt. They bought Riley, their Soft-Coated Wheaton Terrier, when she was ten weeks old. A pedigree, Riley was not considered a plausible show dog because of her gold eyes and brown nose, rather than the required black eyes and black nose.

The Woodleys enjoyed life with Riley, and weren't looking for another dog ... but sometimes fate intervenes, and along came Nolly, believed to be an American Standard Cocker Spaniel.

When she came to the Woodleys Nolly was ten. She'd not had an easy life up until then, and being leash-aggressive might have been part of the reason she'd been abandoned at the age of 7. Found wandering the streets in a nearby city, 'Nolly was adopted by an older person. Although of a very sweet disposition, Nolly was wilful in some ways, and still leash-aggressive. Whilst her owner was in Jamaica for a month, Nolly was boarded at Kinder Kennels, owned by Anna Sanford. On his return, Nolly's owner advised Anna that he was

While Riley and Nolly don't snuggle up, a close friendship and reliance exists between them.

not well enough to take her back, and asked her to try and find Nolly a new home.

Riley also stayed at Kinder Kennels at times, and this is where she first met Nolly: round about the same size and age, the pair got along and became friends. It was also at the kennels that Richard met Nolly, who seemed immediately drawn to him, whilst her warm personality and deep brown eyes soon had Richard enamoured. Anna noticed Nolly's strong attraction for Richard, and suggested he take her home ... he didn't require much persuading!

Laid-back is how the Woodleys describe

Relationships, whether we are talking people or dogs, are sometimes complex, and never black-and-white ... although, in this case, in one sense they are! Things are looking a little strained here ...

Riley, and it's certainly the case that she took co-inhabiting with newly-arrived Nolly in her stride. Nolly is a little bellicose, but this only becomes evident when each of them scrabble to get out the door first when a squirrel goes across the decking. Nolly is food-oriented and, on occasion, there have been some mild exchanges regarding food. "They also sometimes bump into each other when running around, give a non-aggressive growl, and then both back-off," remarked Richard. Riley likes to play, but Nolly is a reluctant participant at best. Yet, if Nolly is ever startled, she will run to Riley to be comforted.

Of course, much of Nolly's character and foibles stem from her earlier life, but she has certainly settled into the Woodley lifestyle and become a member of the family. There is no jealousy for attention or affection between the two dogs; both love their walks, and wait patiently together at the door, ready to set off.

Although the Woodleys have given both Riley and Nolly love and stability, these two little dogs have brought far more into their lives. Said Marian: "They help us focus; provide us with a lot of comfort, along with the responsibilities of taking care of them; keep us active and get us out. All they want is love and we just want them to feel happy and secure."

Zahra and her dynamic duo puppies

Zahra and her two remarkable dogs, Terra, a female Chinese Crested Powderpuff mix, and her adopted brother, Rumi, a Yorkie, divide their time between New York City and country living in Prince Edward County, Ontario. Both were rescue dogs, as was Zahra's original dog, Siggy.

Zahra told me: "I rescued Sig and he literally rescued me right back. He had been found on the street and was severely sick and underweight." Siggy was such a wonderful companion to Zahra that her mother began looking for a dog like him when fate stepped in. As Zahra explained: "I met Rumi at the rescue where I volunteered when he was about nine months old. Rumi melted into my arms when I went to see him, and he and Sig looked like they had been with me forever." Zahra's mother subsequently adopted Rumi as her dog.

Sadly, Siggy unexpectedly died from natural causes. "Siggy was my service dog. I was devastated," said Zahra. "I always told him he shouldn't leave me, but if he had to he should find another dog for me. Siggy was my first dog and my first true love. I became sick after Siggy died, and was advised to get

another service dog, even though I believed that Sig was irreplaceable as he was so much more to me."

Checking various sources, Zahra found a pup who was similar in age and looks to Sig, as well as having the same wonderful expression. Said Zahra: "Terra, the pup, was near San Francisco where Sig and I had lived for a while. Terra was a puppy farm rescue,

Walking is great healthy exercise for us and our dogs, and a stroll in the sunshine helps relieve stress; there's the added benefit of all those conversations with other dog lovers and their dogs!
(Courtesy Zahra)

Life keeps on rolling along, and what better way to enjoy it than with two loving, faithful companions. As Zahra astutely pointed out, "It's also just great to have two to cuddle and play with. I think that's why I have two hands! If two can work in your lifestyle, you will be in for double the joy!" (Courtesy Zahra)

Like Rumi and Terra one should always be fashionably dressed when enjoying the tasty delights at a local NYC café. Zahra noted, "They go to the groomer together. I thought it important that they get used to the touch of other people. I think they give each other courage if something is a bit scary." (Courtesy Zahra)

who almost died in a fire, along with 76 other dogs. I contacted the rescue centre, then flew down to get her. I had not even met her at this point, and was aware that it may not work out. I promised the centre that Terra would have an amazing life, and I was committed to finding her the perfect home, even if it wasn't with me. I had been volunteering with rescues for years by this time, and was very experienced with rescue dogs and helping them with their issues."

Zahra's experience with rescue dogs

Over land, over sea, let's not forget pup safety. While most animal carers are aware of canine safety restraints in cars, canine life jackets are essential if on the water, as these two little sea dogs demonstrate.

(Courtesy Zahra)

paid off: many without her knowledge and patience might have given up on Terra.

It wasn't going to be easy from Zahra's point of view either. "I had a hard time letting Terra into my heart as I was still mourning," she confessed. "She bit me when we first met. I sat on the floor and waited for her to get used to me. Once she seemed to accept me I took her home. She melted in my arms. My doc met her and right away said I had found the perfect service dog. We were inseparable. By this time, Rumi had lived with Zahra's mom in Toronto for

around ten years, but was not well behaved. He would often rip up things in the house, bark loudly and incessantly, growl at people, and be aggressive with other dogs. When Zahra spent time with Rumi in NYC, she would train him, but Rumi resumed his old ways with her mother.

Six or so months later, Zahra's mum decided to relinquish Rumi, and Zahra went to collect him. Rumi and Terra were to meet for the first time on neutral ground, and Zahra described their first encounter. "They got along okay. They both loved me and I had one

All in the family

hand for each pup. We flew to my home in NYC. We got off the elevator to my apartment and the two of them ran to the front door like they owned the place. They walked into the apartment and the tension was palpable."

Once again, thanks to her training, Zahra made sure that both received the same love and treats; any scuffles were immediately intercepted, and she began to teach them to communicate in the best way.

"It was hard with Rumi. He was awful: he simply didn't know how to behave," confessed Zahra. "It took patience, and Terra was a great role model. As a result Rumi learned, and instead of barking when his bowl was empty, he learnt to tap it. He learned he could destroy his toys but not anything else. Most importantly, he learned that people and other dogs are good, too."

Soon the two could be found sleeping in

Zahra, Terra and Rumi split their time between living in the Big Apple and in the country. They all love what the city offers: bright lights, exciting venues, and fun times. (Courtesy Zahra)

Being so photogenic it's easy to see why Terra and Rumi are regularly called upon for print and media ads. Terra's good friend, Shaggy, gets in on the act to create this skateboarding scenario.

(Courtesy Zahra)

the same bed, being protective of each other, and looking for each other if apart. Enthused Zahra: "It's so much fun to wake up with these two. Terra spins and jumps in bed, barks with pure joy, and tries to get Rumi to chase her. And, opening the door and watching them run out together is such a magical moment of our day!"

Dogs, cats - and dragons ...

While the majority of pet owners who opt for more than one pet usually end up with a dog and a cat, this isn't always the case. Take Tamara Des Cotes and her family, for example.

As Tamara explained, "My partner's family had always had Shelties, but I had never before had a pet. We decided to get a Bearded Dragon (Max), as a first pet in preparation for getting a dog, and I fell madly in love with Beardies."

Around two-and-a-half years later, Mikey the Sheltie joined the family. Said Tamara, "Mikey was a curious little puppy, and very interested in Max. He sniffed him so much that Max puffed out his beard at Mikey, though quickly calmed down."

Over time, Tamara found that, typically, Max and Mikey didn't even notice each other very much, to the extent that Mikey often sat on top of Max!

Mikey the Sheltie with Harry and Frankie the Bearded Dragons just hangin' out together.
(Courtesy Enfys Photography)

Mikey tries to entice Max into playing a fun game of ball. "It's a nice ball; great colours." says Max, ... but, no thanks!" (Courtesy Tamara Des Cotes)

Next to join their growing family was another Beardie named Harry, who was completely indifferent to Mikey from the moment they met. For his part, Mickey had grown to be fairly oblivious to these non-aggressive reptiles.

After Max passed away, along came another Beardie – Frankie, who was two-years-old – to join Harry (around six years of age), and Mikey, nearly six also.

Tamara told me Frankie was neither impressed nor bothered by the dog. "He tends to give Mikey an occasional puff of the beard if he gets in his way, but usually just scurries away to let him pass." As far as playing together Tamara said, "Mikey loves to play, while the lizards prefer to snuggle. Mikey would happily play ball with the lizards, but they can be kind of lazy, so playing together typically involves family snuggles and photo shoots."

The Sheltie and Dragons have been the perfect pet combo with Tamara's young child. "All of the animals are incredibly patient with our child, allowing petting sessions, cuddles, and awkward fetch games. I would say the

Mikey consistently doesn't notice the lizards and sits right on top of them. They get even, though by perching on his head! (Courtesy Brittany Boudreau)

Baby Annika cuddles with Max and Harry, who also love to cuddle. Harry, in particular, is a very chill, cuddly guy. Tamara confirmed that he will snuggle with anybody close to him. *(Courtesy Alice Zhao)*

and perspective." stressed Tamara, adding, "Teaching us about respecting different species and their needs has been very rewarding."

Jennifer Cobb, Georgia, Oreo and Mystic

Jennifer Cobb of Wellington, Ontario, is an entrepreneur who shares her home with Georgia, an eighteen-month-old, female Golden Retriever, and two four-year-old male cats, Oreo and Mystic, who are brothers.

Jennifer's two cats arrived with her

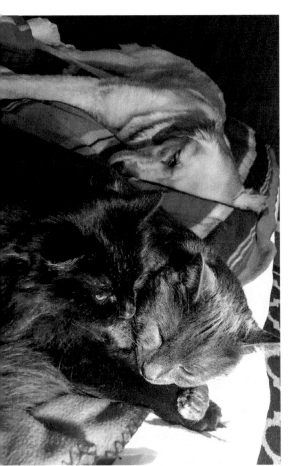

Everybody seems to love a comfortable couch, and nobody has a problem sharing. No tension here; everybody just snoozin'. (Courtesy Jennifer Cobb)

A cat and a dog can play and share together, although Georgia seems to be saying 'Hey, isn't that my ball?! (Courtesy Jennifer Cobb)

pets are more patient with her than they are with each other!" laughed Tamara.

Tamara and her family love having Mikey, Max and Frankie because of the balance they bring. "Beardies are extremely chilled and cuddly; Shelties tend to be a little more energetic and like to be right in the centre of things. They all mean the world to us. No matter what is happening in our life our animals are there beside us to give both love

"Each night before bed I ask all of the animals to sit for a treat. Do you know how long it takes to teach a cat to sit on cue?" laughed Jennifer. (Courtesy Jennifer Cobb)

at different times. Oreo had been abused, and came via a rescue organisation: once in a loving environment, he became very affectionate. Jennifer's neighbour had rescued 27 neglected cats, finding homes for all of them. When one was not working out she begged Jennifer to take Oreo's sibling, which she did. Despite the stable environment it took around three years to gain Mystic's trust, but patience, understanding and love finally won the day.

Mystic has become so attached and grateful now, sleeping alongside Jennifer, and a confirmed 'momma's boy.' The brothers clean each other and love to play-fight.

Golden Retriever Georgia came to Jennifer as a pup, intended to half-soften the blow of Jennifer's daughter leaving home to attend university.

Whilst providing the additional companionship Jennifer sought, having Georgia also opened up conversations with people she never thought she'd meet, and Georgia also made her slow down and 'be there in the moment.'

Relaxing at a village café with her Golden at her feet, Jennifer finds that Georgia's big brown eyes and friendly disposition mean that few can resist stopping to introduce

It's hard to believe that some folk still think cats and dogs don't get along. Yes, they both want to lay in the sunshine, but there's far more going on here. Cuddling is one thing, but hugging shows real affection. (Courtesy Jennifer Cobb)

themselves and their own dog. Being so gentle, parents are eager to introduce their young children to the affection of a pretty puppy.

Overall, though, Jennifer feels that she and Georgia keep each other company.

Jennifer was somewhat apprehensive regarding the cats meeting Georgia, and concerned that the felines should continue to feel safe, since they had already had enough trauma in their lives.

When Georgia was introduced to them, the cats were shocked. "Like, what is *that*?! They all sniffed noses, the cats hissed, and a frightened Georgia crept under a chair," recollected Jennifer.

Happily, that was the worst of it and, after a couple of days, Jennifer was waking up to find all three animals on her bed.

Georgia was pretty well devoid of any natural prey instinct, but Jennifer worked hard with her to tame any hunting/chasing inclination. Georgia never even thinks about chasing the cats, and, in fact, often licks their ears.

While all three eat together, the cats typically pick at their food and leave some of it. That's too much temptation for Georgia, so their dishes are picked up. After they all eat they go out.

"It's the expected routine," Jennifer told me, "They all have a sense of belonging."

Georgia, Oreo and Mystic greet Jennifer when she comes home and when she wakes in the morning. They line up and sit for treats,

and, should she be unwell, sit with her and keep watch.

"It's like we are all part of one pack." Jennifer smiled.

Lee and Lauren Tremblay with Lillee and Sam

Retired professionals Lee and Lauren Tremblay of Picton, Ontario, have had at any one time two, three and four pets for years. And, as far as pet hair is concerned, they consider it a badge of honour.

The Tremblays were ready for two new animal companions, and acquired their male Tabby, Sam, from the local Humane Society, buying a female Golden Retriever they named Lillee at about the same time. Both animals are golden in colour, and both were born in April 2018.

All in the family

The couple decided they wanted two animals for very good reason, as Lauren explained. "We have found that they keep each other company, keep each other busy, and have a partner with whom to conspire against us humans from time to time! The energy level in the home is increased in a positive way. As a retired 'empty nest' couple, they provide great company and purpose. We call our dog our personal trainer: come rain or shine she needs a walk or two every day!"

Sam immediately appeared comfortable in his new home, and was soon joined by 8-week-old Lillee.

Since it was a long drive home after Lillee was collected, she slept the whole trip, and introductions were made the next day.

Sam and Lillee are being raised together from pup and kitten: more brother and sister than cat and dog. (Courtesy Lee and Lauren Tremblay)

Lillee and Sam are never far apart, either sitting, playing, eating or sleeping together ...
(Courtesy Lee and Lauren Tremblay)

Lillee weighed 6lb and Sam just 1½lb. Lauren remarked, "They were both so tiny, but began to prance and play and tear around the room. We kept them in our master bedroom initially, and slowly introduced them to the rest of their new home."

More than a year-and-a-half later Lillee weighs-in at 75lb and Sam 11lb. If Lillee is a bit too playful and annoying, Sam's natural feline agility comes in handy; he simply leaps up onto

... or simply watching the world go by!
(Courtesy Lee and Lauren Tremblay)

Lauren feels that Sam and Lillee, "... think they are the same species. Lillee actually played like a kitten, batting with her paws."
(Courtesy Lee and Lauren Tremblay)

a mantelpiece, or heads to his food room for some relaxing downtime.

Food can often be an issue but is easily dealt with as Lillee eats in the mudroom beside the kitchen and Sam's food is served in the laundry room, blocked off by a baby gate. Sam can easily jump this, but Lillee can't, which is just as well as she loves Sam's food!

Lillee and Sam really love playing together, and often race through the house and wrestle. Apparently, if the play gets a little rough for Sam, he just gives Lillee a swat and walks away!

According to Lauren, "My niece once asked why Sam's head was always wet." I told her, 'Because it is in Lillee's mouth most of the time!"

Despite their size difference, because they grew up together Sam and Lillee get along just fine. "But why does Sam always steal my favourite ball?" wonders Lillee. (Courtesy Lee and Lauren Tremblay)

Still, Sam gets his own back as, once in the fenced yard with Lillee, he's a bit of a scoundrel, running along the outside of the fence to annoy Lillee on the other side.

All in the family

As Lauren noted, "They 'chase' each other back and forth, and Lillee has no chance at all of ever getting Sam."

While life for the Tremblay family is generally thumbs-up, they did have a scare when Sam was about a year old. As Lauren recounted, "Sam went missing for eight days and we thought we had lost him. Then, he showed up at our door, very skinny, literally on his last legs ... Lillee was so gentle and kind. She had missed her friend and knew she could not play with him as she usually did. It took Sam a week to recover and return to his old, naughty self."

Lauren laughed, "They were soon back tearing through the house!"

Lori Campbell & family

The Campbell family's first companion animal was a Japanese breed of dog, a female Shiba Inu. A much-loved, friendly pet, Breckin is 14 years old, and lives with two previously rescued – and equally treasured – Domestic Shorthair male cats: Timon (9) and Peanut (10).

Interestingly, while Breckin has always disliked other dogs, she likes cats.

When Peanut – the family's first cat – joined the Campbell household, he was very timid and very scared, and "... literally would not come out from under the futon in the spare bedroom. After weeks of this, we decided we would have to return him as he was obviously not happy with us." explained Lori. "We decided to get another cat in the meantime; also a rescue. As soon as we brought Timon home, Peanut came out from under the futon, and they immediately became best friends." Problem solved in the best possible way!

Generally, everyone gets along well, playing together and keeping each other

Breckin and Peanut share a couch and a snuggle. (Courtesy Lori Campbell)

company. Breckin likes to bark at one of the cats, who then runs around the house, and the chase is on!

Yet, there are times when they fight like, well, cats and dogs! As Lori explained, "Sometimes they fight – just like human siblings do. We are unsure why they do it, but it happens." These minor skirmishes are soon forgotten, however, as with all families, and grudges are not held.

Often, dog versus cat issues are settled only with a higher degree of subtlety. Lori recalled, "One time the cat was sleeping in the dog's bed, but the dog wanted to sleep in it, so she crawled in, too, gradually inching over until the cat had no choice but to leave the bed as there was no room for him!"

Visit Hubble and Hattie on the web:
www.hubbleandhattie.com • www.hubbleandhattie.blogspot.co.uk • Details of all books
• Special offers • Newsletter • New book news

... and chickens, sheep and ponies!

Melanie's mix-and-match

Melanie Parkes of Queensville, Ontario, has always loved animals and children. A former elementary school teacher, Melanie went on to open her own pre-school out of her country home.

Her acreage was also home to her various animals, who consisted of a pony, a chicken, cats, and a dog. Over the years, as is the norm, sadly, animals aged and died, but new ones arrived, including a lamb, to the delight of her little pupils and her family.

Despite the old cartoon scenario of birds being terrorised by cats, under the right conditions and guidance these two very different animals can become the best of friends. Collie Rusty was also enamoured with Nellie. (Courtesy Melanie Parkes)

Melanie Parkes cuddling her pet red hen, Nellie, with Rusty, her Collie, taking it all in. Rusty was adopted from the local Humane Society.
(Courtesy Melanie Parkes)

Melanie has homed 50-60 cats, with, at one point, 12 at the same time, though is currently down to two.

Jeepers, Melanie's Tabby, likes to cuddle, whilst her black-and-white Shorthair, Collie, is more standoffish. Both like to play with a small ball, batting and chasing it across the floor and around Melanie's feet. Jeepers likes to carry the ball in her mouth and take it to bed at night.

So how did this assorted collection of animals come about?

"Animals tend to come to me," remarked Melanie. "Lost strays 'acquire' me. Once, sitting on our front porch, a number of pigs came walking down the middle of the road. All of a sudden they stopped, came up our driveway, and filed into an open pen behind the garage. We closed the door, and later the police arrived to ask if we had seen any pigs who had escaped from a farm well down the road."

Melanie had no set procedure for introducing the animals to each other. Rescues were isolated for a month after their inoculations and then just simply joined the group.

"Odd," said Melanie, "but these rescues seemed to know they were lucky to be given a home; even the chicken. When our dog, Rusty, met Jesse, a new black cat, they simply sniffed and never bothered each other after that. They accepted the new members. I think a lot has to do with the people concerned, though."

Melanie found that having the animals get along was easy, with a routine for meals, rest, treats and play, and great enjoyment of petting and grooming. "Everyone also stakes out their favourite chair, window, etc."

Her first pony was Rascal, who was joined by the equally lovable Polka-Dot, a Shetland/Welsh pony mare, now the most senior resident at 34 years of age. Rascal and Polka-Dot became very close, and when Rascal died, Polka-Dot refused to go into her stall.

Melanie tried everything, and was at her wits end when she learned of an Animal Communicator. On the telephone the Animal Communicator told Melanie that Rascal was still very much present in spirit. Recalled Melanie, "She said to tell Polka-Dot that Rascal wanted her to go in the stall: that Rascal needed her to do it." After doing this, amazingly, Polka-Dot turned and went into her stall without fuss from that point on.

Some time later Polka-Dot was asked to share her small stable with a male lamb called Domino. Domino, now 8, and Polka-Dot are now inseparable. About the only time Domino leaves Polka-Dot's side is when he hears the sound of the bird seed bouncing off the concrete near the back door. An unusual treat – and one of the few things that Polka-Dot doesn't have an interest in – she neighs loudly whilst Domino's away, calling him back to her.

As a tiny lamb Domino lived indoors – more like a puppy – with Scooter the part Border Collie/Lab. As he got bigger, Domino adopted a bigger sister in Polka-Dot the pony. (Courtesy Melanie Parkes)

Feeling safe and secure is what every animal wants. As long-time residents, Polka-Dot and Domino wander freely while Melanie is outside. Obedient as any dog, Polka-Dot and Domino come when called; especially if it means a treat. (Courtesy Melanie Parkes)

Surrounded by more than an acre of tasty grass and with Melanie (and treats!) nearby, it's a great life, and a safe haven. (Courtesy Melanie Parkes)

For Melanie, animals really are all in the family as one of her sisters has many dogs and another birds. As far as Melanie is concerned, "People could learn a lot from animals. They are full of love."

Cindy and Rick Rogers: a continually growing family!

Cindy and Rick Rogers are professionals in health and education fields respectively, and residents of Oakwood, Ontario.

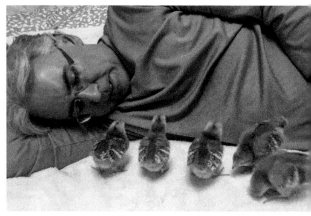

Rick meets the latest additions to the Rogers family: baby Buckeye chicks, who appear to be more fascinated by than frightened of this giant lying in front of them. (Courtesy Rick & Cindy Rogers)

Separation anxiety can exist even between two very different domestic animals, such as Polka-Dot the pony and Domino the sheep. Whilst Polka-Dot neighs for Domino to return to her, he seems to be saying: "Stop neighing at me! I'm eating my bird seed as fast as I can and I'll be back in a minute!" (Courtesy Melanie Parkes)

Here's Newton the older male Australian Shepherd, and the svelte male Potcake (Turks and Caicos Island dog) Danny. (Courtesy Rick & Cindy Rogers)

Moving from the city where they had a few companion animals – canaries, turtles, and a couple of cats – the Rogers relocated to the country, and "thought we should add a dog, for security and as a companion."

They ended up with two: Newton is a 13-year-old male Australian Shepherd, and Danny, a male Potcake (Turks and Caicos Island dog) half his age. Interestingly, Rick was allergic to cats whilst growing up, while Cindy always had them. "We finally did have a couple of cats and he was okay," recalled Cindy, and once ensconced in the country, "animals seemed to find us!"

At an old abandoned dairy barn at the end of their road, Rick found some very sick kittens on top of a manure pile (in fact, there were over 25 cats and kittens). "We couldn't walk away," remembered Cindy. "We paid to have some spayed/neutered, and then, via the Toronto Humane Society, became involved in a trap, neuter and return program. We eventually got all of the cats sterilised, and then either adopted or returned them to the Society for adoption.

"All of the cats are chipped in the tip of the left ear and tattooed to signify they have been through the program. There were many incidents during the trapping, from mistakenly catching raccoons to having to lay food late at night in order to catch these kitties."

Not surprisingly, many of their current cats are rescues from the barn, including Tea, Tori, and Jessie, plus Bella, who they kept as she was a surrogate mother to a litter of kittens from the barn. Others included Cassie who was found wandering in the forest across the road, while Panda came from another barn, found suffering with a gangrenous tail.

Cindy noted, "We adopted Miss B as a one-eyed kitten from China. We had been told by an animal rescuer that the Chinese tend not to adopt animals unless they are perfect. Another rescuer asked for our help. We were concerned that if Miss B needed surgery down the road to close up the empty eye socket, her adopters would not want to pay for it.

"Miss B and our cat, Mia, got along so well, that we kept her: a year later Miss B needed that very expensive surgery."

Kenzie was a foster cat who never left. She was one of 30 felines at a hoarder's house, and had suffered a severe case of frostbite to her toes, plus most of her left ear was missing.

Said Cindy, "She was so scared when she came here as she was not used to human contact. Eventually, she bonded with us."

Mika was born in the Rogers' spare bedroom. "We fostered her mother, Crystal, and she had four kittens: many kittens have been born in our spare room."

While cats and dogs are traditional pets, some aren't quite as usual. Cindy recalled, "We started with chickens after another nurse where I work told me about hers, and how easy and fun they are to care for. We had read about the cruelty that is rife in the egg

Crystal and her kittens snuggle up whilst Mika (right), whom the Rogers adopted, joins the family group.
(Courtesy Rick & Cindy Rogers)

industry, and thought it would be great to have fresh eggs, and also educate others. Of course, we realised when we saw what personalities they have that we could not continue to eat chicken, and have been vegetarian ever since, with vegan dishes, too."

The Rogers care for Goldie, Splash, Molly and Rose, who are all beautiful female Gold Laced Wyandottes. They also enjoy the company of four Buckeyes named Matilda, Stasia, Paige and Tina (aka Tiny).

Introducing new members to an ever-growing family can sometimes be tricky, although there was never an issue with introducing Newton the Australian Shepherd to the cats as he was petrified of them! Danny, the Potcake dog (whose name comes from the congealed peas and rice mixture traditionally eaten in the Caribbean, as the overcooked rice that cakes to the bottom of the pot would be fed to the dogs) took a while, as he would snap playfully at them. Claws in the snout put a stop to that.

Said Cindy, "Newton was a natural with the chickens as he was afraid of them, too! Danny took a while with the chickens as he would try to charge at them through the fence.

"We began letting the chickens out to free range, and Danny learnt to keep watch over them," recounted Cindy. *"He now goes into their fenced area to act as a guard dog. Sadly, we lost one of our young chicks to a red-tailed hawk this summer, so Danny being there gives us peace of mind."*
(Courtesy Rick & Cindy Rogers)

Mia closely observes an unwell Goldie. Animals are very sensitive and perceptive, and many would agree that there is concern in Mia's eyes, knowing her friend is very ill. (Courtesy Rick & Cindy Rogers)

"Bringing chickens in for a bath was always entertaining," chuckled Cindy. "We weren't sure at first how the cats would react, but Mia was fascinated by them and had to show up to help with the bath." (Courtesy Rick & Cindy Rogers)

We were afraid to trust him with them as he would chase birds on the beach in Turks and Caicos, where he came from."

As for the cats, Mia met the chickens one day while on a harness.

Cindy recalled, "Not sure who was more afraid: her or the chickens. Mia would not give up trying to get into the spare bedroom where we raised our four Buckeye chicks from a day old. She was fascinated with them and never tried to hurt them, although we never left them unattended as we never knew if her natural instincts might kick in."

Overall, the Rogers have found that the dogs tend to play with each other, and the cats do the same. Although Mia has tried to play with the chickens, apparently, they were not impressed by this idea.

All in all, there have not been any really serious concerns or reactions amongst all of the diverse members of the Rogers family.

Cindy feels it's "... amazing having the unconditional love of so many animals. I have been battling some health issues this year and they are often my motivation to keep going."

An added bonus is that their animals have brought new friendships into the couple's lives. Cindy explained, "I started an Instagram account [@8coolcats] to post about my wacky household of animals, and have met many cat, dog and chicken people as a result. A gentleman called Mike, who lives in Washington State, has chickens, cats, dogs, and tortoises, and, since meeting, now has indoor Seramas, the world's smallest chickens."

Over time, Cindy and Mike became good friends, and he and his wife flew to meet her

Cindy summed up her animal family in these terms: "They are all our furry and feathered children, so to speak. When the universe didn't bless us with children of our own we thought we could open our hearts and home to the many animals who just seemed to find us!" (Courtesy Rick & Cindy Rogers)

and her family. "Amazing that our chickens connected us across the miles " laughed Cindy.

Janice and Gary: hair, fur and feather!

Janice and Gary are semi-retired, and live in Prince Edward County, but keep themselves busy with their assortment of animals.

Long-time animal lovers, on more than one occasion the pair have crossed the domestic line and taken in creatures from the

wilder side of life. Currently, every one of their 7 animals has been a rescue, except for the two Cockatiels.

Their two male Shelties – Gus, a 10-year-old Standard Tri, and 8-year-old Max, a Blue Merle – were adopted from a breeder. Gus was a puppy when adopted, whereas Max was already six years old when he was adopted. Apparently, the two get along famously as long as Max remembers that Gus is the boss ...

Dogs have ways of conveying what they are thinking. Barking is the most obvious, but sitting in front of the fridge is a subtle reminder that it's time to eat ... (Courtesy Janice Windsor& Gary Gamble)

The black male kitten, Simba, arrived at their door on Thanksgiving night a year ago. He was hungry and tired, but very friendly. Whereas Gus had grown up with cats, Max had his doubts about the newcomer, but that certainly wasn't the case with Simba, as he followed on behind the dogs like one of the pack (much to Max's chagrin).

Moving on to the couple's feathered

While Max is a little apprehensive of Simba, fellow Sheltie Gus has no qualms about helping Simba guard the stairs.

(Courtesy Janice Windsor & Gary Gamble)

Simba loves teasing the dogs with one of his toys. It used to be that all of the toys belonged to Gus, but, since the cat moved in, they're all now in his box!

(Courtesy Janice Windsor& Gary Gamble)

population, Teal, known now as T'ielk, is a 15- to 20-year-old South American Conure (parakeet), whose newly-divorced Toronto owner no longer wanted.

Said Janice, "T'ielk – who was living as part of a large family of assorted tropical birds – was left behind with 'Mom' when the couple divorced, but she was afraid of him – for good reason – and didn't want to keep him. The marital problems of his original owners caused T'ielk to lose some of his feathers – and others he ripped out himself. Being left behind when all of his feathered family departed meant more stress and subsequent feather loss for poor T'ielk. He still chews on his shoulders (out of habit as much as anything), and an avian vet has told us that this won't change now.

Just goes to show how what social creatures birds are, and how external influences can adversely affect them: so sad.

"We changed his name as he reminded me of the off-planet warrior king from a sci-fi television series."

The two male Cockatiels are father and son, Said Janice, "Having female birds is a problem as they keep wanting to lay

Cockatiels Billy and Mork are easy to please with regard to food. Said Janice, "As long as they get their millet strips, they're happy."
(Courtesy Janice Windsor & Gary Gamble)

Pedra is constantly on the move, too, and almost impossible to photograph, except for this one in her house cage, when she actually came over and posed! (Courtesy Janice Windsor & Gary Gamble)

eggs, which often results in premature death." [Many become egg-bound.]

A pigeon rounds out their current bird assortment. Originally, 'Pedro' was thought to be a male, but it was soon discovered that he was really a she, so her name was changed to Pedra.

As often happens, Pedra was an unplanned addition. As Janice explained, "Pedra was rescued from a barn that was being cleared out. She had no feathers, no family, and was going to be abandoned, so we took her in. Unfortunately, we can't let her go now as she hasn't been taught how to survive in the wild. She would probably walk up to a hawk or fox and ask for a cuddle."

"Ya, okay I've lost a few feathers, but I'm still a pretty boy!" Janice commented, "I had a difficult time getting T'ielk in a shot as he hates the camera, and starts running around his cage when he sees it. You have to sneak up on him to get a picture."
(Courtesy Janice Windsor & Gary Gamble)

As previously mentioned, introducing new pets can be tricky, especially if it is an altogether new species. Whilst introducing different breeds of dogs to each other was a simple process, adding Simba to the mix was carefully planned. A clean bill of health and the onset of colder weather meant it was time to bring Simba inside the house, but in a restricted area.

Said Janice, "We blocked off the laundry room so that the dogs could see and smell Simba, but couldn't get to him. This setup continued for a couple of weeks until we were sure that there wasn't going to be conflict between the animals; then we allowed Simba to wander through the house. We suspect there must have been dogs in his previous residence as he was completely comfortable. In fact, he follows them around constantly."

The birds are housed in cages in a separate room filled with plants, although had arrived at different times. Janice described the living arrangements thus: "The dogs, as a general rule, are indifferent to the birds, and we let them in to interact. The two Cockatiels could not care less as they had been around dogs. Pedra the pigeon appears to like dogs, and would probably enjoy riding on one of them if allowed to. She doesn't appear to have any fear of them, but we don't want to test that idea, just in case."

Apparently, T'ielk the Conure doesn't much like anybody, including Janice and Gary at times.

Whilst everybody likes to socialise (with the exception of T'ielk), the two older dogs aren't keen on playing. Young Simba the cat is, though, and constantly wants to jump out and scare the dogs, or try and get them to chase him.

And then there's the special relationship between T'ielk and Pedra. Janice explained, "T'ielk does say a few words, albeit they are somewhat garbled. He makes a huge fuss over Pedra, especially when we bring her in from her flight cage outside, and screams, 'Pretty girl!' then 'Pretty boy!' and carries on until somebody responds and lets him know that Pedra is in for the rest of the day, and he is indeed a pretty boy. He usually struts around the cage at that point, showing-off to Pedra, then begins muttering under his breath – swearing, I think!"

T'ielk is shy when asked a question. Janice recalled, "I often walk back into the room and ask if everyone is okay. Surprisingly, one day, T'ielk answered me, saying, 'Alright, we're okay, Bob.' We've no idea who Bob is, and were afraid to ask!"

Although Pedra has no fear of the dogs or cat, she is far from fearless, as Janice discovered.

"One day I heard a great commotion coming from the pigeon's outdoor flight cage. I went to investigate and found that a Monarch butterfly was fluttering around the cage. Pedra was absolutely terrified. I had to go into the cage and convince the butterfly to leave so that Pedra could calm down."

Janice had to admit that living with a multitude of animals can be a challenge at times, especially when one wants to be in charge and another doesn't want to co-operate. Still, she conceded, they usually figure it out.

Janice and Gary concluded with, "There are a million benefits of having animals. Their undying love, attention, and trust is incomparable. These are all our babies and we couldn't give up any of them!"

John, Terry Lynn, Jacob and Hannah Phillips & their allsorts animals

Professionals John and Terry Lynn Phillips, their children and additional family members, live in Sutton West, Ontario.

John had a long association with animals before he married Terry Lynn; she says, "John had always had more than one animal. John's grandparents had a kennel in Toronto (Parker Pet Care), which is now owned by his aunt and cousins. John's grandmother bred Standard Poodles, and, as a young boy, he spent much time in the kennel, eating kibble and playing with the animals.

"As John grew up, the family had several dogs and cats at the same time, but also had a raccoon (Penelope) who adopted them, living both inside and outside the house, and a Canada Goose that had been injured and became domesticated."

In contrast, Terry Lynn had only a fish and a dog, a much-loved Maltese who died after just a couple of years, sadly.

When Terry Lynn first met John his family had a parrot called Walter, Ralph, a German Shepherd, and Reg, a black Lab, plus Captain, a 3-legged cat, and two ferrets. His sister also showed horses, and kept several of them in the barn.

Marriage is a time of great 'adjustment' in two individuals' lives, and Terry Lynn admitted that this meant "taking on a collection of animals." Having several companion animals at the same time was not always a conscious decision on Terry Lynn's part, but she and John have had multiple animals on occasion since marrying, and, at the very least, have always had a dog.

Their extended family has consisted of Earl the crow, whom they believe now roosts in the trees around their house; a 7-year-old male cat called Murphy, who hit the road one day in search of his fortune, and a now deceased African Grey parrot called Walter. Other animal family members have included a long-haired guinea pig, two female ferrets, and another dog and cat.

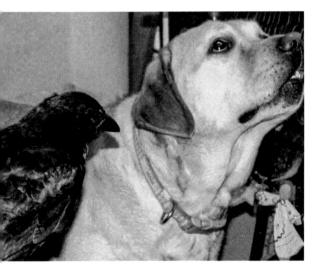

Terry Lynn recalled fondly, "Earl the crow loved Nellie our Lab. Earl and Nellie would play fetch, and just lounge together. When Earl was around Nellie was never too far away." Earl liked to watch hockey sitting on John's shoulder, and his favourite game was when the Chicago Black Hawks played the Detroit Red Wings. (I'm sure he was a Wings fan!)
(Courtesy the Phillips family)

Currently resident are a rescued Chinchilla named Freddie, and the matriarch of the family, 13-year-old Nellie, a yellow Labrador Retriever. Some of the creatures were chosen; some rescued, and Earl, a nestling, was saved when a tree crashed in their yard and dumped out the nest he was in.

Hand-fed and kept inside in his fledgling days, Earl adopted the Phillips as his alternative family, and fully integrated into their lives and, ultimately, their neighbours', too.

As Labs are known to do, Nellie calmly took everything and everybody in her stride, including Earl. "Earl grew up thinking he was a dog, I swear, and even made a noise that sounded like barking." laughed Terry Lynn. "He became a sidekick to Nellie."

Earl is not a fan of the cat, Murphy, and always stays on higher ground when he is around.

Then there's Freddie the Chinchilla, who was apparently pretty hyper until he got comfortable with the zoo. He would scurry

"Earl gave us many amazing moments. He was always taking things, like keys. He took the keys to the ride-on lawn mower while they were still in it, and we could never keep a key outside. Our daughter, Hannah, was locked out once by Earl's trickery," recounted Terry Lynn. Earl would also bring yellow Lab Nellie little sticks as presents. Here, John tries to convince Nellie to impress on Earl not to try to take the keys to his truck.
(Courtesy the Phillips family)

"You and I look alike, but what big ears you have, Freddie!" says Murphy. "All the better to hear you with when you try sneaking up on me!" replies Freddie. Although skittish at times, Terry Lynn said the Chinchilla would let her clean his cage – including using the vacuum – whilst he was still in it! (Courtesy the Phillips family)

"And then there was the night that Murphy somehow lifted the latch on Freddie's cage, and John found him inside the cage with Freddie," sighed Terry Lynn. "Freddie was not very happy with sharing his cage. There was a lot of running back and forth." (Courtesy the Phillips family)

around his cage in a bit of a frenzy, and the Phillips also discovered he absolutely hated men, which may have stemmed from past experiences.

As Terry Lynn observed, "Our cat, Murphy, spent a lot of time tormenting Freddie by going up to his cage and trying to engage. Murphy liked to lounge about and sleep on top of Freddie's cage. Eventually, they made friends, but only as long as Freddie was in his cage." Freddie did escape once; apparently, it was like "having a squirrel ripping around the house."

Murphy had no issues about dogs, and, in fact, loved playing cat games with Nellie.

Terry Lynn summed up life with her extended family with, "Different species, like different people, should be allowed their differences, but live peacefully with each other. The benefits of having several other species

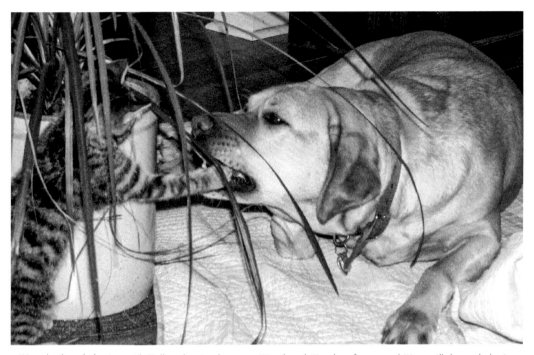

"Murphy loved playing with Nellie, chasing her, swatting her, biting her face as a kitten will do, and playing hide-and-seek. Nellie was very tolerant, and if she'd had enough she would simply walk away." remarked Terry Lynn. (Courtesy the Phillips family)

living with you is that life is never boring; for you, or for the animals. There is always something interesting happening, and, in our case, certain animals were able to have a home where they were wanted and looked after properly. They enriched our lives, and enabled us to pass on a love of animals to our children."

"In summer when we ate outside on the deck, Earl would sit on our son's head, whilst Nellie watched carefully for anything that might fall on the ground." Continued Terry Lynn, "The experience with Earl was particularly rewarding and educational. Today, our son, Jake, is particularly calm and comfortable with animals." (Courtesy the Phillips family)

Family album

Why lie on a hard floor
when there's a comfy
sofa to hand ...?
(Courtesy G & G E Donaldson)

"Sorry, but this sofa's taken ..."
(Courtesy Wes & Glenda Meyer)

Lucky by name, lucky in life! This much-loved, 200-pound pig lives a pampered life indoors with his carers and his friend, the cat. When introducing a new animal into your home, ensure that the existing residents are happy about the new arrangement, as well as the newbie.

"This is my friend's little Cockatiel, Phoenix, and my Umbrella Cockatoo, Niles," said owner Mary Orr. Despite the difference in size, the Cockatiel tried to bully Niles by biting at his toes.
Mary went on to say, "Niles kept looking down at him in disbelief as if (in my imagination) to say, 'Hey, you're gettin' a little irritating there, son ...'"
Alas, not everybody gets along with everybody else. Just like people ...
(Courtesy Mary Orr)

A younger Samantha Moumouris regularly sings to her 30-year-old Conure parrot, Amber, who loves to be serenaded. The Moumouris family found that, with animals around, there was always someone singing. Even visitors would sing to the bird. Certainly, Amber is totally absorbed in the soft, sweet melody sung by Samantha – as is Sawyer the Collie.
(Courtesy the Moumouris family)

Male Bearded Dragon Bruce Lee poses for the female Naudica. Unfortunately, most of these creatures don't live very long – usually five to 8 years. As Janice Windsor noted, "They have special needs and diets, but they are really cool and quite affectionate." (Courtesy Janice Windsor & Gary Gamble)

All animals (including people) suffer from varying degrees of separation anxiety. As an integral part of the family, our companion animals want to be involved as much as possible in all that we do. Often just being there with the family is enough, but sometimes – and with dogs especially – they want to join in the fun, too!

Musically inclined? Birds can make you an opera star, or at least help you along the way. Said Janice Winsor, "I learned to whistle and can now actually carry a tune reasonably well, which is something I could never do prior to having birds." (Courtesy Janice Windsor & Gary Gamble)

"You know, Macallan, I feel so much more secure when wearing these proper safety harnesses whilst riding in the Morgan." "You're right, Whinnie, I am sure glad our owners bought that excellent book Dogs on Wheels, by Norm Mort! (Courtesy G & G E Donaldson)

Dogs are like kids ... they love a game of rough-and-tumble in the snow. People not so much, but it's heart-warming to watch the dogs having so much fun. *(Courtesy Jennifer Cobb)*

Patrick has a typical Beagle personality: loving, loyal and rather boisterous. Yet, even a great watchdog must rest at times, and a favourite spot is on the couch, which he somewhat begrudgingly shares with the cat!
(Courtesy Cheryl Douglas)

Sharing and patience are valuable traits in animals, too. Nellie the Lab knows her friend, Earl the Crow, will leave some of that watermelon for her. "Boy, that looks delicious!"
(Courtesy the Phillips family)

Training is always an ongoing process. Zahra advised that the latest training for her pups centred on car journeys. "Rumi gets nervous. I think he has motion sickness, so constant training is required to help him overcome this. Terra is perfect in the car, but gets trampled on when Rumi goes nutty. They would share a car booster seat, but Terra would be on edge – literally! I recently added a pillow beside the car seat so Terra could have her own space. That has been working out splendidly. Of course, when the car is in motion, both dogs are safely secured by their seat belt harnesses." (Courtesy Zahra)

Other great books from our Hubble & Hattie Kids! imprint

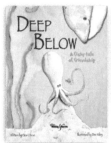

9781787117464

9781787117389

9781787117198

9781787117792

9781787117488

9781787117372

9781787117730

9781787116993

9781787115163

9781787113060

9781787113121

9781787111608

9781787112926

9781787115156

9781787113077

9781787114302

9781787117631

9781787113862

9781787114180

www.hubbleandhattie.com/hubbleandhattiekids/